Jet Flying Boats

DAVID OLIVER

AMBERLEY

First published 2018

Amberley Publishing
The Hill, Stroud,
Gloucestershire, GL5 4EP

www.amberley-books.com

ISBN 978 1 4456 4613 8 (print)
ISBN 978 1 4456 4614 5 (ebook)

British Library Cataloguing in Publication Data.
A catalogue record for this book is available from the British Library.

Typeset in 11pt on 14pt Celeste.
Origination by Amberley Publishing.
Printed in the UK.

Contents

Introduction

During six years of the Second World War, more than 9,000 flying boats were produced by all the main protagonists, fulfilling a multitude of roles including maritime reconnaissance, bombing, fleet spotting, search and rescue, long-range transport and communications.

At the end of the war, more than 1,000 flying boats and amphibians remained in military service. However, time was running out, and a little over a decade later the military flying boat would appear to be a dying breed on the verge of extinction.

With the termination of hostilities in 1945 came the cancellation of contracts for many advanced and innovative flying boat designs.

At the same time, post-war commercial aircraft development moved away from the flying boat, which had been used to pioneer long-haul air travel in the 1930s when Imperial Airways Short flying boats were a key factor in linking up Britain's worldwide empire, and Pan American Clippers carried passengers across the Atlantic and Pacific oceans.

Due to the fact that the reliability, range and economy of landplanes had improved dramatically during the war years, and that by 1945 there was an abundance of airports with long, paved runways all over the globe, again products of wartime expansion, the commercial flying boat, which was difficult to maintain and required expensive marine terminals, quickly lost favour with airlines, and they rapidly became regarded by them as relics of a bygone age. The case for the commercial flying boat was not helped by the fact that few leading aircraft manufacturers were willing to risk the high costs of developing new types, using up-to-date technology for a dwindling civil market, or without extensive government backing for military versions. Therefore, only civil conversions of wartime designs, such as the Short Sunderland, were available to compete with the latest breed of post-war landplane airliners.

It would be another twenty years before a new breed of multi-engine flying boats would leave the drawing boards of leading aerospace companies to successfully fulfil not only their traditional military roles of maritime reconnaissance and search and rescue, but one at which they have proved to be very adept – aerial firefighting.

The flying boat's resurgence has been prompted by the availability of reliable, powerful and economic turboprop and turbofan engines. In the meantime, there have been many attempts to produce a successful modern flying boat that have ended in failure as a result of either technology or politics.

The beginning of the end for British post-war flying boat production – the first and last jet- and turboprop-powered British flying boats are seen here at Cowes in 1952: Saunders-Roe's SR./A1 is on the left and the SR45 Princess on the right. (Richard Riding Collection)

Modern flying boats have a niche role as aerial firefighters, examples of which are the turboprop-powered Canadair CL-415, on the left, with the jet-powered Beriev Be-200 dominating the scene. (Andrew Salnikov/Beriev)

CHAPTER ONE

Saunders-Roe's Swan Song

Following the cancellation of contracts for many British flying boat designs, one that almost survived was the Saunders-Roe SR.A/1. Developed during the Second World War by the pre-war Empire flying boat designer Sir Arthur Gouge, the SR.A/1 was a single-seat jet fighter flying boat. Powered by two 3,230 lb thrust Metro Vickers Beryl turbojet engines, and armed with four 20 mm Hispano Mk 5 cannons, it was also designed to carry two 1,000 lb bombs or eight rockets on underwing hardpoints. The SR.A/1 was also the first fighter aircraft to be fitted with a Martin-Baker ejection seat.

The British fighter flying boat was the first of its type since the Japanese Mitsubishi F1M Pete, the Nakajima A6M2-N Rufe and the Kawasaki N1K Rex aircraft of the Second World War.

With a single oval intake, the two side-by-side engines were fitted into the roots of the shoulder-mounted wings and were fitted with individual exhausts angled out by 5 degrees to ensure the efflux cleared the hull. A high-mounted tail plane avoided jet efflux and spray. The SR.A/1 had inward-folding wing floats and a water-rudder could be attached to the main rudder when on the water.

The Ministry of Supply ordered three prototypes with the first, TG263, flying at Cowes on 16 July 1947 with Saunders-Roe's chief test pilot Geoffrey Tyson at the controls. Despite an all-up weight of 16 tons higher than many contemporary land-based fighters, the SR.A/1 had an excellent rate of climb and a maximum speed of over 500 mph. Unfortunately, the second and third SR.A/1s were lost during flight trials. The third aircraft, TG371, which first flew on 17 August 1948 and was fitted with the more powerful 3,850 lb thrust Beryl turbojets, was being flown by legendary test pilot Lt-Com. Eric 'Winkle' Brown, who had achieved Mach 0.82 in a dive before the aircraft struck a floating object while landing at Cowes. Although the pilot managed to jump clear, the aircraft sank without trace. Only a month later, the second SR.A/1 crashed into the North Sea while being flown by Squadron Leader Major, who was rehearsing aerobatics for an air show.

However, Geoffrey Tyson had already shown the flying boat fighter's aerobatic capabilities at the 1948 Farnborough SBAC Show when he finished his slot with a low, inverted pass over the runway.

By 1949 Saunders-Roe had studied the possibility of re-engining the SR.A/1 with a single Sapphire 3 turbojet, which was being developed by Armstrong Siddeley, who had acquired the rights to the Metrovick Beryl programme. However, with no funds available, the one remaining SR.A/1 was mothballed. Despite this, the outbreak of the Korean War in June 1950 stimulated some activity by Saunders-Roe's marketing team, who saw the possibility of deploying the flying boat fighter from depot ships at sea or from rivers and lakes. By this time its performance had been outpaced by land-based jet fighter aircraft and the project was abandoned, though the SR.A/1 was de-mothballed to make its final public appearance on the River Thames during the 1951 Festival of Britain, which, ironically, celebrated British industry, arts and science.

Although the SR.A/1 had demonstrated Saunders-Roe's innovative and adventurous approach to aircraft aerodynamic and hydrodynamic design, the SR.A/1's protracted development and lack of investment was due, in part, to Saunders-Roe's concentration on another complicated and challenging flying boat programme

Even prior to the Second World War's end, the British Air Ministry was enthusiastic at the prospect of reviving the pre-war Empire flying boat routes. The Ministry approached the two leading British manufacturers of metal-hulled flying boats, Saunders-Roe and Shorts, requesting that the two firms collaborate on the development and manufacture of a new aircraft, which would emerge as the Short Shetland. Saunders-Roe designed the flying boat and manufactured the wing while Shorts produced the remainder of the aircraft. Two examples of the Shetland were completed, but the first military prototype was destroyed by fire at the Marine Aircraft Experimental Establishment (MAEE) at Felixstowe, and although a second civil aircraft was flown in 1947, the project was considered a failure and it was scrapped in 1951.

During 1943, the chief designers of Saunders-Roe and Short had collaborated to produce a preliminary design specification for an innovative large flying boat. This design specification defined various criteria for the proposed aircraft, which would be the largest all-metal flying boat ever built. Performance criteria included a weight of 140 tons, a pressurised 'double-bubble' hull, a 214-foot wingspan, a height of 56 feet and a length of 148 feet. The proposed aircraft would be capable of a cruising speed of 340 mph at an altitude of 37,000 feet, with a range, dependent on payload, of between 3,640 to 5,100 miles, all with luxurious accommodation for a total of 104 passengers.

In May 1946, the bid made by Saunders-Roe was selected as the winner, leading to the company receiving a £2.8 million contract from the Ministry of Supply for the production of three SR.45 flying boats, later to be named the Princess. In 1948 the construction of the first SR.45 Princess commenced at the Saunders-Roe works at East Cowes.

The 330-ton flying boat was powered by ten Bristol Proteus turboprop engines arranged as coupled pairs and two outboard single engines. The Proteus was designed to develop 3,500 shp and de Havilland-designed contra-rotating four-blade propellers were fitted to the coupled engines.

The launch of the Princess into the Cowes waters was delayed until the night of 20/21 August 1952 due to adverse weather. However, weather conditions on 22 August were perfect and the prototype, G-ALUN, conducted its maiden flight, piloted by Geoffrey Tyson. The initial flight lasted for thirty-five minutes, in which the flying boat performed a complete circumnavigation of the Isle of Wight.

However, what was not known at the time was that the Princess programme was already dead in the water. Hopes that the flying boat would be taken into service with the British Overseas Airways Corporation (BOAC), the post-war successor to Imperial Airways, were dashed when it was announced that in 1949 post-war flying boat operations would be run down. This decision was not rescinded and by November 1950 all BOAC flying boat operations had ceased. This left Saunders-Roe without its most important potential customer.

Conflicting views surrounded the future use of the Princess for long-range passenger services, but despite ceasing flying boat operations, BOAC continued to have an interest in the project. As late as 1951 this interest led to the formation of the BOAC Princess Unit to prepare the three flying boats under construction for their introduction into service. The principal duties of the unit included dealing with plans for provision of engineering equipment and other technical facilities, carrying out route surveys, the provision of economic and financial information about the development and operation of the aircraft, the development of special operational methods and procedures and the development of BOAC's Hythe flying boat base for use by the new aircraft. Despite the work carried out by the Princess Unit, it was apparent by mid-1951 that the airline's interest in the flying boat was on the wane.

Although the BOAC chairman, Sir Miles Thomas, witnessed the first flight of the Princess, in October 1952 he expressed the view that in his opinion the Princess was out of date technically.

The prototype was quickly put through several test flights with the hope that the flying boat could be sufficiently readied to appear at the 1952 Farnborough SBAC. However, indications of engine reliability issues were encountered and this led to the intended flight display at Farnborough being abandoned.

In 1951, BOAC had undertaken an in-depth re-evaluation of its standing requirements, and determined that the airline presently had no need for the Princess, or any new large flying boat. The airline had already elected to terminate its existing flying boat services, but the Ministry of Supply announced that construction of the three Princesses would proceed with the intention of using them as transport aircraft in RAF service to carry up to 230 fully equipped troops. However, in March 1952 it was announced that, while the first prototype would be completed, work on the second and third Princesses would be suspended to await more powerful engines.

Meanwhile, as further testing of the prototype continued, the reliability of the engines and gearboxes continued to be problematic, but not to the extent that flight testing was prevented. The Princess was subject to handling trials on behalf of the MAEE at Felixstowe and received a favourable report. Evaluation of the flying boat continued into 1953, during which particular attention was devoted towards addressing the difficulties uncovered. During the 1953 Farnborough Airshow, the prototype was displayed with daily low-level passes. Flight tests of the prototype continued up until 27 May 1954, by which point it had been found that the Proteus engine, once perfected, would be capable of enabling the type to achieve its projected performance figures. G-ALUN had performed forty-six test flights in total, during which 100 flight hours were accumulated.

The Princess Air Transport Co. Ltd was formed with the objective of studying the factors affecting the operation of the Princess flying boats, and to tender for their operation, should the opportunity have arisen. Meanwhile, the two unflown Princesses were stored at the former RAF flying boat base at Calshot.

In late 1953, Aquila Airways, the only British airline operating flying boats, offered £3 million to purchase the Princesses, but the government turned down the offer. Interest in the aircraft was shown by the US authorities in 1956 to convert the Princesses to fly either the General Electric Direct Air Cycle or Pratt & Whitney NJ-2B nuclear-powered jet powerplants as part of the United States Army Air Force (USAAF) Aircraft Nuclear Propulsion (ANP) programme. Studies were undertaken and tank test models of the Princess were carried out by both Convair in San Diego and the Glenn L. Martin Company in Baltimore. However, USAAF and US Navy interest in the project diminished and nothing came of the venture.

In 1958 a plan was revealed to operate the three flying boats on services between Southampton and Canada and Rio de Janeiro in Brazil. The backers of this plan included a businessman, B. G. Halpin, and the former managing director of British South American Airways, Air Vice-Marshal Don Bennett. The plan included re-engining the Princesses with six Rolls-Royce Tyne 11 turboprops in place of the ten Proteus engines, but after waiting more than a year for a decision from the Ministry of Supply, the idea was abandoned.

In 1964 the three Princesses were purchased by Eoin Mekie on behalf of Aero Spacelines, which planned to use them as heavy-duty freight aircraft for transporting Saturn V rocket components for NASA. However, when the cocooning was removed, it was found that the aircraft were badly corroded as the contract for maintenance and inspection of the stored aircraft had been allowed to lapse, which resulted in the rapid deterioration of the airframes. Consequently, the deal was terminated. All three Princesses were broken up by 1967.

While the fate of the Princesses was being sealed, future developments of jet-powered flying boats were being studied by the Saunders-Roe design team under Henry Knowler. The Duchess design was to be powered by six de Havilland Ghost turbojet engines installed in the wing section. Accommodation was planned for seventy-four passengers in two cabins. The 130-ton flying boat would have a cruising speed of 500 mph and a range with maximum payload of 2,600 miles. The Duchess design offered a similar performance to the contemporary DH Comet jet airliner with the operational flexibility of the flying boat.

In 1956 the Saunders-Roe P192 proposal was submitted to the P&O Company as a giant cruiser of the air. With a 313-foot wingspan and a projected all-up weight of 680 tons, and being powered by twenty-four Rolls-Royce Conway 1 turbofan engines, the P192 could carry 1,000 passengers in luxurious accommodation on four decks.

The P.208 was a twin-engine flying boat Maritime Patrol Aircraft (MPA) pitched in 1958 and, due to NATO requirements, built around advanced submarine detection equipment. Powered by two Rolls-Royce Tyne 11 turboprop engines, the 150-foot wingspan aircraft had a design all-up weight of 73,000 lb. A retractable hydro-ski was built into the hull, boosting take-off from the water. A sixty-five-passenger civil version was also considered.

However, nothing came of these ambitious projects and the Princesses were the last fixed-wing commercial aircraft to be produced by the Saunders-Roe Company.

The Saunders-Roe SR.A/1 prototype jet fighter flying boat 'on the step' as it takes off from Cowes, with Geoffrey Tyson at the controls. (Francois Prins Collection)

This air-to-air view of the Saunders-Roe SR.A/1 shows the flying boat's hull, jet exhausts and retracted wing floats. (Francois Prins Collection)

The Saunders-Roe SR.A/1 in flight shows its clean lines, high-mounted tailplane and clear bubble canopy. (Francois Prins Collection)

The third SR.A/1, TG271, being prepared for a test flight. This aircraft was lost when piloted by Lt-Com. 'Winkle' Brown after striking a floating object on landing on 12 August 1949. (Richard Riding Collection)

The SR.A/1 made its final public appearance on the River Thames during the 1951 Festival of Britain. (Author's collection)

The first and only surviving SR.A/1 is seen on its original beaching gear, demonstrating its metal cockpit hood with small windows. (David Oliver)

The hull of the first prototype Saunders-Roe Princess flying boat, seen under construction at the Columbine Works in East Cowes. (Richard Riding Collection)

Technicians checking the complex Bristol Proteus turboprop engines in preparation for the first flight of the Princess give scale to its immense size. (Richard Riding Collection)

The first Princess flying boat is seen carrying out engine and taxi trials at Cowes in August 1952. (San Diego Museum)

Seen during its thirty-five-minute maiden flight on 22 August 1952, the Saunders-Roe Princess flew around the Isle of Wight. (Richard Riding Collection)

Princess G-ALUN was given an attractive airline-style paint scheme during 1953 to cover its bare aluminium finish. (Richard Riding Collection)

A stillborn Saunders-Roe flying boat proposal was for a four-jet, seventy-four-seat Dutchess, which was designed to compete with contemporary airliners such as the DH Comet. (Author's collection)

CHAPTER TWO

Convair's Lost Causes

At the height of the Cold War, the Korean War had sparked renewed interest in the flying boat, particularly in the United States, where several new advanced designs appeared in the mid-1950s – the most exiting of which was another jet fighter flying boat, the supersonic Convair Sea Dart. Having watched the development of the British SR.A/1 with interest, the US Navy embarked on the Sea Dart programme, which would push the state-of-the-art aerodynamics and hydrodynamics to the limit.

The Model 2 Sea Dart was designed to meet a requirement for a flying boat interceptor capable of Mach 0.95, but was expected to achieve Mach 1.5. Convair's proposal gained an order for two prototypes in late 1951. Twelve production aircraft were ordered on 28 August 1952, before a prototype had even flown. No armament was ever fitted to any Sea Dart, but the plan was to arm the production aircraft with four 20 mm Colt Mk 12 cannon and a battery of folding-fin unguided rockets. Four of the contracted aircraft were re-designated as service test vehicles, and an additional eight production aircraft were soon ordered as well.

A slim, thin, delta-wing single-seat aircraft was built, utilising a blended hull and a retractable double hydro-ski undercarriage, and the pressurised cockpit near the end of the pointed nose was equipped with an ejection seat.

The Sea Dart was designed to be powered by two 6,100 lb thrust Westinghouse XJ46-WE-02 turbojet engines. However, these were not available and the prototype was fitted with two 4,080 lb thrust afterburning Westinghouse J34-WE-32 turbojets instead.

The aircraft was built in Convair's San Diego facility at Lindbergh Field and was taken to San Diego Bay for testing in great secrecy in December 1952. On 14 January 1953, with E. D. 'Sam' Shannon at the controls, the aircraft inadvertently made its first short flight during what was supposed to be a fast taxi run, although its official maiden flight was made on 9 April 1953.

The underpowered engines made the fighter sluggish, and the hydro-skis were not as successful as hoped. They created violent vibration during take-off and landing, despite the shock-absorbing oleo legs they were extended on. The two skis were replaced by a single ski that improved this problem, but Convair was unable to resolve the aircraft's sluggish performance. The Sea Dart proved incapable of supersonic speed in level flight with the J34 engines, and its pre-area rule shape caused higher transonic drag.

The second XF2Y-1 prototype was cancelled, so the first of four YF2Y-1 service test aircraft was built and flown with the J34 engine, later to be retrofitted with the J46-WE-02s.

Although the aircraft suffered from poor visibility due to its nose-up attitude and lack of power on take-off, flight trials continued without mishap for two and a half years, during which time it became the first, and only, flying boat to break the sound barrier when the YF2Y-1 exceeded Mach 1.0 in a shallow dive on 3 August 1954. The aircraft proved to be extremely manoeuvrable on the water, with direction when water taxiing being controlled by the water rudder and asymmetrical engine power. Take-off runs of twenty-five seconds over a stretch of 2,300 feet of water were achieved and successful landings in Sea State 5 were carried out with no adverse stress on the airframe.

However, on 4 November 1954, the first YF2Y-1 disintegrated in mid-air at 5,000 feet over San Diego Bay during a demonstration for naval officials and the press, killing Convair test pilot Charles E. Richbourg when he inadvertently exceeded the airframe's limitations.

Even before that the US Navy had been losing interest due to problems with supersonic fighters on carrier decks having been overcome, and the crash relegated the Sea Dart programme to experimental status. All production aircraft were cancelled, including the projected single-engine XF2Y-2, although the remaining three service test examples were completed. The final two prototypes never flew. Limited trials continued until 1957 but the full performance envelope was never explored. The Sea Darts logged over 300 flights in forty-six months before the US Navy discontinued the programme in 1957.

Convair, which produced one of the most successful Second World War flying boats, the Catalina, had also been developing a very different flying boat at the same time as the Sea Dart – the first turboprop-powered four-engine long-range transport flying boat, the Tradewind. Derived from the XP5Y-1, a prototype reconnaissance bomber took off from the San Diego bay on 18 April 1950, being flown by Convair test pilot Don Germeraad and with Sam Shannon as co-pilot. In August 1950 the XP5Y-1, powered by four 5,250 shp Allison XT-40-A-4 turboprops, set a world endurance record in its class with a flight of eight hours and six minutes.

However, the US Navy announced in the same month that it no longer had any need for an armed patrol flying boat and the gun installations on both sides of the fuselage and the tail were removed. In fact, the planned five pairs of 20 mm guns were never fitted. The XP5Y-1 was lost on 15 July 1953 on its forty-second flight when the elevator torque tube broke and the aircraft could no longer be kept under control. It crashed into San Diego Bay after the whole crew had left the plane by parachute. A second XP5Y-1 never flew and was finally scrapped in 1957.

With a long, graceful hull with a tall, narrow fin and rudder, the Tradewind's straight-tapered wings, mounted above the fuselage with four high-set engines, had non-retractable V-braced wingtip floats.

A small number of Tradewinds were produced for the US Navy in two versions: the R3Y-1 for personnel transport or ambulance duties, and the R3Y-2 for vehicles or heavy freight.

Both had a 145-foot wingspan and were powered by four 5,850 shp Allison T40-A-10 turboprop engines. These contained a double-coupled powerplant with extension shafts and a reduction gearbox for the contra-rotating propellers. At maximum emergency power it could produce 5,100 shp at a weight of only 2,495 lb, including the drive shafts. It could also give a maximum continuous power of 4,000 shp. However, the drive shaft and gearbox system was far from reliable and the T40 could only be used for fifty hours before overhaul was necessary. Since it was also used in a few other aircraft types, great efforts were made to improve the reliability of the engine. Without substantial success, it was finally only produced in modest numbers for some experimental aircraft types.

The first flight of the R3Y-1 was made on 25 February 1954. Prior to its launch on 17 December 1953, it was christened by the then leading Hollywood star Esther Williams with a bottle of water collected from the seven ocean seas as Tradewind *Indian Ocean*. Six of the US Navy Tradewinds were also christened with ocean and sea names.

The fully pressurised R3Y-1 was 142 feet long and could accommodate a crew of seven, plus 103 troops, or ninety-two stretcher patients and twelve medical attendants, while the 139-foot-long R3Y-2, which made its first flight on 22 October 1954, was designed as a flying Landing Tank Ship (LTS) to taxi up to a beachhead and disembark four 155 mm howitzers, or six jeeps, or assault troops using built-in ramps under a bow-loading door that hinged upwards. One of the R3Y-2 aircraft set a transcontinental seaplane speed record of 403 mph by utilising the jet stream – a record that still stands today.

Another R3Y-2 was used for flying tanker trials in 1956 with four refuelling pods under its wings. These carried drogue-type hoses that permitted the simultaneous refuelling of four fighter aircraft. Fuel could be transferred at a rate of 950 litres per hour (l/hour) and the total weight of the in-flight refuelling equipment was about 1,000 lb, which could be installed in less than five hours. The trials, refuelling four US Navy Grumman F9F-8 Cougar jet fighters, and later four US Navy McDonnell F2H-4 Banshee fighters, were a great success, and the Tradewinds were cleared for use as tankers. In spite of these successful trials, the Tradewind was never used operationally as a tanker.

A total of eleven Tradewinds, with a maximum speed of nearly 400 mph and a range of 4,500 miles at 300 mph, were operated by Naval Air Transport Squadron

2 (VR-2) of the US Navy's Fleet Logistic Wing, which was based at Naval Air Station Alameda, California, from 1954.

However, in January 1958, Tradewind *Indian Ocean*'s number two engine's propeller sheared off, puncturing the fuselage en route from Honolulu to San Francisco. It was lost while attempting an emergency landing, which the crew survived. Tradewind *Coral Sea* had crashed earlier, also due to a propeller malfunction.

Both of these incidents hastened the end of the operational career of the Tradewind, and on 16 April 1958 they were all decommissioned. A year later they were all scrapped. The Tradewind never fully met the expectations for a reliable multi-purpose transport flying boat. Although it gave demonstrations in beach landings and in air-to-air refuelling, the Tradewind remained in US Navy service for less than three years, having been plagued with persistent engine maintenance problems. With the modest numbers built and its relative low flying hours, the fleet of eleven Tradewinds accumulated only 3,300 flying hours in total, and as such it must be regarded as a failure. As with the British Princess flying boat, it was a case of the slow pace of the development of powerful and reliable turboprop engines that ultimately sealed their fates.

The failures of Convair's last flying boat projects mirrored those of Saunders-Roe and the Sea Dart, and the Tradewinds were Convair's last flying boats to take to the air.

The first prototype of Convair's advanced jet fighter flying boat, the XF2Y-1 Delta Dart, taxies out into San Diego Bay under its own power. (Convair)

Above: The delta-wing single-seat XF2Y-1 Delta Dart is seen on its unique twin hydro skis during a high-speed run during sea trials in 1953. (Convair)

Left: The Convair XF2Y-1 Delta Dart was the fastest flying boat ever built when it broke the sound barrier in August 1954. (Convair)

Able to come ashore under its own power, the XF2Y-1 taxied up the Convair seaplane ramp on its twin hydro-skis. (RAF Museum)

Taxiing out of the water and onto dry land required a considerable amount of thrust from its two Westinghouse XJ46 turbojets. (US Navy)

Due to porpoising problems with the skis, the XF2Y-1 was fitted with a single retractable hydro-ski undercarriage for improved stability. (Convair)

One of the three surviving Convair YF2Y-1 Sea Darts is now on display at the San Diego Air & Space Museum. (MAP)

The Convair XP5Y-1, a prototype reconnaissance bomber powered by four Allison XT-40 turboprops, flew for the first time on 18 April 1950. (US Navy)

The Convair XP5Y-1, designed to be armed with five pairs of 20 mm guns, crashed into San Diego Bay on 15 July 1953 on its forty-second flight. (RAF Museum)

An R3Y-1, the long-range troop transport variant of the Convair Tradewind, taxies into San Diego Bay during its early trials. (Convair)

The second Convair R3Y-1 is rolled out on its beaching gear at the Convair factory at San Diego. (US Navy)

The graceful lines of the R3Y-1, which was the world's most powerful flying boat when it entered US Navy service in 1954. (Convair)

Powered by four 5,850 shp Allison T40-A-10 turboprop engines, the R3Y-1 could carry 103 troops over a 4,500-mile range. (Convair)

The Convair R3Y-2 could carry howitzers, military vehicles and supplies, or an assault company of US Marines on its 88-foot-long cargo deck. (Convair)

The bow-door R3Y-2 was also used for flying tanker trials in 1956 with four refuelling pods under its wings to air-refuel four US Navy Cougar jet fighters. (Convair)

After successful air-refuelling trials with US Navy Banshee fighters, R3Y-2 Tradewinds were cleared for use as tankers. However, they were never used operationally. (Convair)

CHAPTER THREE

Martin's Magnificent Failure

At the end of the Second World War, more than 1,000 flying boats and amphibians remained in military service around the world, and the largest fleet belonged to the US Navy. But time was running out, and a little more than a decade later the military flying boat would appear to be a dying breed on the verge of extinction. However, before the end of its era, a number of advanced, but ultimately unsuccessful, designs were launched, harnessing the latest technology available at the time – namely, the jet engine.

One of the most exciting of the new post-war generation of flying boats was another US Navy programme of the era that almost succeeded – the Martin SeaMaster, which was the most advanced flying boat to ever be produced in the United States.

In the immediate post-war era, as the Cold War warmed up following the Berlin Airlift and the outbreak of the Korean War, the US Air Force Strategic Air Command (SAC) was the United States' global peacekeeping force by virtue of it being the sole means of delivery of the nation's nuclear arsenal. As a consequence, SAC had more than 1,000 nuclear bomber aircraft by the mid-1950s, most of which were jet-powered B-47s and which comprised most of the US defence budget.

The US Navy saw its strategic role being eclipsed by the USAF and knew that both its prestige and budgets were at stake. Its first attempt having been a victim of budget cuts – the USS *United States,* a large 'super carrier' intended for the launching of US Navy strategic bombers – it chose instead to create a Seaplane Striking Force (SSF), which was useful for both nuclear and conventional warfare, including reconnaissance and minelaying. Groups of these aircraft, supported by seaplane tenders or even special submarines, could be located closer to the enemy, and, as mobile targets with no fixed air bases, they would be more difficult to neutralise.

The US Navy issued a request to industry in April 1951. The SSF seaplane was to be able to carry a 30,000 lb payload to a target over 1,500 miles from the seaplane's aquatic base. The aircraft was also to be capable of a Mach 0.9 dash at low altitude. Convair and Martin submitted proposals, with Martin winning the competition.

The Glenn L. Martin Company, named after its founder, was established in 1909 and would go on to produce many successful large flying boats, starting in 1935 with the graceful M130 four-engine long-range civil flying boat. With a 130-foot wingspan, three China Clippers were used by Pan American Airways to pioneer its transpacific routes – two of which were impressed for war service with the US Navy. A single larger version, the M156, was built in 1937 and later sold to the Soviet Union. During the Second World War, the Martin Company produced more than 1,000 twin-engine, gull-winged, medium-range maritime reconnaissance bombers – the PBM-1 Mariner – and five four-engine, long-range naval transport planes – the M170 Mars, which at the time of its first flight in 1942 was the world's largest flying boat. Both of these types would see service in the Korean War, when they were joined by another Martin product that would become the US Navy's last operational flying boat when it retired from front line service in 1966 – the impressive twin-piston-engine P-5M Marlin maritime patrol aircraft.

On 31 October 1952, the US Navy awarded Martin a contract for two SSF seaplane prototypes – designated M275 and named the SeaMaster – plus a static test article. This initial order soon led to further contracts for six pre-production service evaluation machines, with the designation YP6M-1, and up to twenty-four full-production machines, with the designation P6M-2.

The Martin design team, led by George Trimble, an aeronautical engineer and head of the Martin advanced design department, developed a revised hull design, with a length-to-beam ratio of 15:1, which was felt to offer the best efficiency in both the air and on water. The prototype XP5M-1 Marlin flying boat was rebuilt to test the new hull design for the SeaMaster, with the test aircraft being designated the Martin M270.

The SeaMaster was the world's first multi-jet-engined flying boat that was designed as an anti-submarine aircraft, a minelayer and a nuclear bomber. The innovative design had many novel features, including the high length-to-beam ratio hull formed by reducing the frontal area that was largely responsible for putting the flying boat back into competition with landplanes (at least as far as speed was concerned). It was originally designed to be powered by four jet engines and a Curtiss-Wright compound turbo/ramjet was initially designated as the SeaMaster powerplant. After several failures in development testing, the decision was made to fit the flying boat with four Allison J71-A-4 turbojet engines, which had 13,000 lb of afterburning thrust each, and which were mounted in nacelles in pairs above the wing, near the roots. The J71 was a derivative of the J35 axial-flow turbojet, which was used to power the Republic F-84 Thunderjet and was originally developed by General Electric as the TG-180 but passed on to Allison for full production.

The graceful shoulder-mounted swept wing had a marked anhedral, bringing the wingtips so close to the water that tip floats were attached directly to the wingtips like auxiliary fuel tanks. The SeaMaster had the same variable-incidence 'flying T' tail and spoiler ailerons as the Martin XB-51 – the prototype three-jet, low-level, ground attack landplane that was developed for the USAAF and first flew in 1949. The 'T' tail was designed to avoid spray when taking off and landing on water and the bow snout of the hull was fitted with a spray deflecting strake. Aft of the step, the long and deep 'V'-formed hull reduced pitching when alighting on rough water and improved the take-off. Martin hydroflaps mounted at the extreme rear of the hull bottom acted as an air brake when opened together, and as water rudders when used individually on the water.

A crew of four – the pilot, co-pilot, navigator/radio operator and the flight engineer – was accommodated in a pressurised cabin that incorporated an airlock, enabling a crew member to access the bomb bay and equipment in an unpressurised section of the hull. Aft of the cabin was the weapons bay, which could be loaded through a large hatch in the top of the hull. Up to 30,000 lbs of mines or bombs were carried on a rotating bomb bay door that was made watertight by inflatable rubber seals. The sole defensive armament comprised twin 20 mm cannon in a radar-directed turret in the extreme tail, with tail warning radar and sighting equipment also included.

Much thought went into making the SeaMaster capable of operating without reliance on a fixed base. The top of the engine nacelles were hinged to enable complete engine changes while afloat and special equipment for refuelling from submarines at sea was developed along with special floating launching and beaching vehicles.

The first XP6M-1 prototype made a successful first flight on 14 July 1955 with Martin test pilot George Rodney at the controls, and the second flew on 18 May 1956, but within eighteen months both had been lost in accidents. The test programme had continued smoothly until 7 December 1955, two days after the death of Glenn L. Martin. During a routine check flight for the first US Navy pilot, the first XP5M-1 prototype crashed into the Potomac River, killing all four aircrew. The post-mortem revealed a control-system fault that caused the aircraft to pitch nose down, bending its wings down and ripping them off. The second SeaMaster prototype was refitted with new flight instrumentation and ejection seats, and test flights finally resumed in May 1956.

Later in the year, solutions were being sought for a frequent airframe vibration that plagued both prototypes. One solution involved locking the elevators together with the variable-incidence flying tail. A test flight was carried out on 9 November 1956 to verify the improvement in the vibration. However, in recovering from a shallow dive at high speed, pilot Bob Turner lost pitch control of the second XP6M-1, which started a violent outside loop. The crew ejected safely as the airframe broke up and crashed near Odessa in Delaware. Information from the flight data recorders indicated that the modified tail configuration had been overpowered by dynamic forces at high speed due to a previously undiscovered mathematical error in calculating loads for the hydraulic control actuators.

Nevertheless, the first of six pre-production YP6M-1 SeaMasters flew in January 1958 and twenty-four production P6M-2 aircraft were ordered for the US Navy. These were powered by four 17,200 lb trust Pratt & Whitney J75 twin-flow axial compressor turbojets, which were designed to give a maximum speed of over 633 mph at 40,000 feet and a range of nearly 3,000 miles.

Despite the loss of both prototypes, the US Navy still remained enthusiastic about the SeaMaster. A beaching cradle was designed by the Aeronca Corporation to allow the flying boat to taxi in and out of the water under its own power, and two landing ship docks, two seaplane tenders and the submarine USS *Guavina* were sent to shipyards to refit them as SeaMaster support vessels. A home base was set up at Naval Air Station Harvey Point, near Elizabeth City, North Carolina.

The first pre-production YP6M-1 was rolled out in November 1957, with flight tests resuming in January 1958. It featured afterburning Allison J71-A-6 engines, which were visibly toed out to reduce the effect of exhaust blast on the rear fuselage. The engine inlets were also moved back from the leading edge of the wing, presumably to reduce water ingestion. Five more YP6M-1s were built in 1958 and participated in an extensive flight test programme. Mine-laying and navigation systems were implemented, even though standard US Navy mines could not yet withstand sea impact when dropped at high speed. Successful practice drops from the rotary bomb bay of conventional and dummy nuclear munitions were conducted.

Early in 1959, production P6M-2s began to emerge from the Martin plant, and the full potential of the design was realised. The installation of newly developed 17,500 lb thrust afterburning Pratt & Whitney J75-P-2 turbojet engines gave the P6M-2s nearly 12,000 more pounds of static thrust. This allowed the gross weight to be increased to 195,000 lb from the 171,000 lb of the YP6M-1s. Increased weight meant a greater draft for the hull, which in turn necessitated raising the wing anhedral to zero degrees. Other improvements included full-visibility canopies, with large overhead panels for a better field of view, and transistorised Sperry navigation and bombing systems. Production P6M-2s were to be equipped with mid-air refuelling probes, and a 'buddy-pack' probe and drogue-refuelling kits were designed to fit inside the bomb bays, allowing for their fast conversion into air tankers.

Pilots reported that the SeaMaster handled well and was capable of flying Mach 0.89 'on the deck'. This was important, as the development of radar-guided surface-to-air missiles had made low-level flying an essential part of strategic penetration missions. The flying boat's wings were especially strong for the extra stress of high speeds through thick air, the aluminium skin at the wing roots being an inch thick. By contrast, the US Air Force's contemporary B-47 bomber could manage about Mach 0.58 at low altitude, while the SAC's newer B-52 could only manage 0.55.

During the development of the flying boat, the Seaplane Tender USS *Albemarle* was earmarked for conversion in February 1956 to tender Martin P6M SeaMaster jet flying boats. The ship was reassigned from the Atlantic Reserve Fleet to the Commandant, 4th Naval District, for conversion, effective 6 February 1956. Equipped with stern ramps and servicing booms to handle the SeaMaster, as well as a semi-sheltered area and a service dry dock, the ship emerged from the conversion possessing the capability to serve as a highly mobile seadrome, capable of supporting jet seaplanes anywhere. USS *Albemarle* was recommissioned on 21 October 1957 and, after being fitting out, she sailed for Norfolk in December. The ship then sailed for Guantanamo Bay on 3 January 1958, to carry out shakedown training, before dropping down to Montego Bay, Jamaica. USS *Albemarle* continued to operate out of Norfolk through 1959 and into 1960, although the cancellation of the SeaMaster programme meant that the ship would never service the aircraft for which she had been reconfigured.

Rising costs of the programme, however, had led to two cutbacks, which reduced the number of production P6M-2s from twenty-four to eighteen. Then, by the end of 1958, orders were reduced to just eight. Three production P6M-2s had been completed by the summer of 1959, and five more were under construction. When it was discovered that the cost of these eight SeaMasters would be almost $450 million, the US Navy cancelled the programme completely on 21 August 1959. There were loud protests since the flying boat had performed well, but the SeaMaster was considered an obsolete concept, and the US Navy was already moving full steam ahead to a much more effective nuclear deterrent capability in the form of the Polaris ballistic missile submarine. All the SeaMasters that had been built sat idle for over a year and were then scrapped.

Martin tried to promote other seaplane designs, such as an eight-engine transport version of the SeaMaster that was informally called the Project M307 SeaMistress. Aerodynamic and hydrodynamic tests were made of a 500,000 lb high-subsonic multi-jet logistics transport seaplane design, which conformed to the transonic area rule. The aerodynamic results showed that acceptable stability and performance characteristics could be obtained on a high-subsonic flying boat. Reasonable lift-drag ratios could also be obtained of about Mach 0.90. Additional improvements in lift-drag ratio and longitudinal stability characteristics could be obtained by small refinements in the area distribution and the hydrodynamic behaviour of this design was determined to be generally satisfactory. Preliminary tests indicated that afterbody suction forces introduced some longitudinal take-off instability and high-speed resistance great enough to preclude take-off without afterburning. However, the addition of a small auxiliary step to the afterbody, slightly off the main step, improved the stability and reduced the resistance to the point where satisfactory take-offs could be made without afterburning.

The 218-foot wingspan could accommodate 415 fully equipped troops or 130 tons of cargo that could be loaded by a Tradewind-type bow door that hinged upwards. Powered by eight J75 turbojets, the SeaMistress had a design maximum speed of 650 mph and a range of 1,500 miles.

However, neither the US Navy nor the USAF showed any interest in the project and Martin's long tradition of producing flying boats was at an end. Martin later formally abandoned developing and producing aircraft altogether in order to focus on missile and defence electronics production.

The first prototype XP6M-1 SeaMaster was rolled out at the Martin factory in Baltimore on 21 December 1954. It then made its first flight on 14 July 1955. (Glenn L. Martin Maryland Aviation Museum)

The first XP6M-1 SeaMaster taxiing into the water under its own power on its beaching gear, which was manufactured by the Aeronca Corporation. (Glenn L. Martin Maryland Aviation Museum)

The first XP6M-1 SeaMaster climbs out of the sea and up the ramp under its own power, showing its distinctive anhedral wing and tip floats. (Glenn L. Martin Maryland Aviation Museum)

The US Navy publicly announced the SeaMaster in November 1955 when it invited the press to witness the rollout of the second XP6M-1 prototype. (Glenn L. Martin Maryland Aviation Museum)

The XP6M-1 Sea Master taxies at high speed in Chesapeake Bay in September 1955, illustrating the spray pattern under the 'T' tailplane. (Glenn L. Martin Maryland Aviation Museum)

The graceful lines of the XP6M-1 in flight, showing overwing-mounted turbojet engines and hydro-flaps, which operated as air brakes and water rudders. (Glenn L. Martin Maryland Aviation Museum)

The four pre-production YP6M-1 SeaMasters on the Strawberry Point ramp at Middle River, Baltimore. The flying boat's beaching gear is seen in the foreground. (Glenn L. Martin Maryland Aviation Museum)

Prior to P6M-2 No. 9's first flight in February 1959, the aircraft is seen on the step during a high-speed taxi run across Chesapeake Bay. (Glenn L. Martin Maryland Aviation Museum)

The flight engineer monitors the engine start of a P6M-2 SeaMaster on its beaching gear during pre-flight checks. (Glenn L. Martin Maryland Aviation Museum)

One of the pre-production YP6M-1 Sea Masters taxies ashore under its own power on the inflatable Aeronca beaching gear. (Glenn L. Martin Maryland Aviation Museum)

Powered by four 17,200 lb trust Pratt & Whitney J75-P-2 turbojets, and fitted with a revised clear canopy and an air-refuelling probe, a PM6-2 is seen on its take-off run across Chesapeake Bay. (Glenn L. Martin Maryland Aviation Museum)

CHAPTER FOUR

Mallows to Mermaids

Unlike many NATO countries, the Soviet Union continued to develop long-range flying boats into the twenty-first century. Georgii Mikhailovich Beriev established his experimental design bureau (OKB) at Taganrog, located on the shores of the Sea of Azov, which has been continuously producing flying boats for more than eighty years.

Having relied on Lend-Lease license-built Consolidated PBY Catalinas until the end of the Second World War, the Soviet Navy ordered a new twin-piston-engine, gull-winged, long-range flying boat, the Be-6 in 1949. The 108-foot wingspan flying boat had a crew of eight, could also carry up to forty fully equipped troops, and had a maximum speed of 234 mph and a range of nearly 3,000 miles. The problem-free first flight took place at Taganrog in February 1949 with OKB's chief test pilot, M. I. Tsepilov, at the controls. More than 200 Be-6s were built for the Soviet Navy and Aeroflot, and subsequently for service with the Chinese People's Navy, some of which were re-engined with Wopen WJ-6 turboprops and were designated Qing-6.

In 1946 Beriev explored the possibility of modifying the Be-6 design with two RD-45 turbojets – Soviet copies of the Rolls-Royce Nene – and was sufficiently impressed to request permission to design a smaller jet-powered flying boat, which was designated R-1 and retained the Be-6 gull wing to keep the engines as high as possible from the water. The R-1 had a crew of three, with the pilot in a fighter-type cockpit, the navigator/bomb-aimer in a glazed nose and the rear gunner sighting the twin-NR-23 cannon barbette through blisters on each side of the rear fuselage. Powered by Klimov VK-1 engines, the R-1 made its maiden flight from Taganrog on 30 May 1949. Following a series of successful water landings and take-offs, it was seriously damaged in a landing accident, after which it sank in shallow water. After a comprehensive rebuild, test flights of the R-1 resumed in July 1953, with improved wing slats and a redesigned drainage channel to prevent water seeping into the under hull area having been implemented. A new swept tailplane was designed but only a mock-up was constructed.

The modified R-1 performed very well with no further problems being experienced during take-off and when landing. It continued to serve as a flying test bed and flying boat pilot trainer until February 1956, when test pilot M. Vlasenko performed an emergency landing in the shallow waters of Gelendzhik Bay, which damaged the hull on the sea floor. The R-1 was not repaired.

In response to a 1953 Soviet Naval Aviation (AV-MF) requirement for a replacement for the Be-6, Beriev used its experience of flying an experimental R-1 twin-jet-powered flying boat to design a large swept-wing, twin-jet flying boat with an impressive performance that rivalled many Soviet landplane reconnaissance bombers of the era. With more than a passing resemblance to the Martin SeaMaster, the 94-foot wingspan, 100-foot-long prototype was flown by V. Kuryachi from Gelenzhik on 20 July 1956.

The Be-10 maritime reconnaissance bomber – or 'Mallow', as codenamed by NATO – carried a crew of three. Powered by two shoulder-mounted 14,330 lb thrust Lyul'ka AL-7PB turbojets, it established an impressive number of FAI-recognised records for water-borne aircraft, all of which were set in 1961. These included an impressive speed record over a straight course of 566.9 mph, a 544.2 mph speed record with an 11,000 lb payload over a 1,000 km closed-circuit course, and altitude records of 49.135 feet with 11,000 lb, and 39,360 feet with 33,000 lb.

Its main wing was designed with a marked anhedral with wingtip floats, again mirroring the SeaMaster, although it had a conventional tailplane, unlike the Martin flying boat's high 'T' tail. The pilot, again under a fighter-style canopy, and the navigator/bomb-aimer were accommodated in a pressurised compartment, both with upward ejection seats, while the radio operator/rear gunner, operating the twin NR-23 cannon barbette, had a pressurised cabin with a downward ejection seat. Offensive loads included Kennel air-breathing anti-shipping missiles, homing torpedoes, or depth bombs on under-wing pylons, and additional weapons on a rotary mine door.

However, despite its record-breaking performance, the Be-10 was limited in range and weapons capacity and could only operate in relatively calm waters. Only twenty-seven production aircraft were built, which were issued to the 1st and 2nd Squadrons of the 977th OMDRAP, which became the only operators of the Be-10, operating from a naval base at Lake Donuzlav on the Crimean peninsula as part of the Black Sea Fleet. One of its claims to fame came in the aftermath of the Cuban Missile Crisis in October 1962, when Soviet Navy 'Mallows' flew reconnaissance missions over Turkey to verify that United States missiles had been withdrawn.

By the mid-1960s, both its predecessor, the Be-6, and the Be-10 were in the process of being replaced by the Be-12, a gull-winged, twin-turboprop-powered amphibian that was to become one of the most successful and long-lived flying boats to be built since the Second World War.

By the end of the 1950s, flying boat development in the West, both military and commercial, had virtually come to a halt. However, this was not the case in the

Soviet Union, and Western observers were somewhat taken aback when no fewer than two new multi-engine military flying boats made an unexpected appearance at the 1961 Aviation Day Flypast at Tushino Airport, Moscow. Both aircraft were products of the Soviet design bureau which was headed by the world's most prolific post-war flying boat designer, Georgii Mikhailovich Beriev.

The aircraft that made its debut in the 1961 Tushino display was almost ignored by the Western press, which described it as a turboprop conversion of the Be-6 'Madge'. This description, however, was far from accurate, although it would be another six years before the type was shown to Western observers in detail at the USSR's 50th Anniversary Air Display at Moscow's Domodedovo Airport in 1967, where three production Beriev Be-12s made an appearance. It was then revealed that the Be-12 was a completely new aircraft. As the first amphibian designed by the Beriev bureau to be put into production at Taganrog, it had flown for the first time in February 1959 and entered service with the Soviet AV-MF in 1964.

The overall configuration of the Be-12 Seagull – NATO codenamed 'Mail' – was at first sight similar to its piston-engine predecessor with the familiar narrow high-gull-wing and twin-fin dihedral tailplane. Non-retractable floats were attached to a single streamlined pylon and mounted at the elbow of each wing was a 4,190 shp Ivchenko AI-20D turboprop engine. The compact but spacious hull housed a crew of five or six, including the pilot, co-pilot, radio operator, electronics operator and one or two ASW sensor operators, according to the mission. The Be-12's glazed observation, weapons aiming and navigation station in the nose also housed its radar 'proboscis'. The relatively simple (for an amphibian) single-wheel units folded neatly upwards into the fuselage sides, in line with the leading edge of the wing, while the tail wheel retracted into the rear stern beneath the magnetic anomaly detection (MAD) stinger, and was closed in by two doors. These doors acted as water rudders, although there was a small skeg at the end of the hull.

The traditional defensive gun turrets on most Beriev flying boats were replaced by an observation astrodome in the dorsal position and a MAD extension protruded fifteen feet from the rear fuselage. The sharp planing surface was almost identical to that of the Be-10, although it was considerably shorter in overall length, and included a shallower single step than that of the jet boat. A Be-10-type bow dam was also fitted to the amphibian, which had a lip fitted to the edge of the hull that acted as an additional spray suppressor.

The Be-12 was capable of carrying a total weapons load in excess of 1,500 lb, including torpedoes, depth charges and sonobuoys that were carried in an internal unpressurised bay with watertight doors to the rear of the step, and bombs, guided anti-submarine missiles or rockets that were carried on two underwing pylons outboard of the engines. Sonobuoys, target markers and flares could be launched at sea level through a number of unpressurised launch tubes. The amphibian was able to alight on calm water to search with its own on-board sonar equipment to supplement the air-dropped sonobuoys, and weapons and various stores could be loaded via side hatches in the rear of the hull while the Be-12 was on the water. An auxiliary power unit was fitted beneath the training edge of the port wing, the exhaust of which could be used to heat the leading edge of the tailplane. It was comprehensively equipped with navaids and no fewer than four radar altimeters.

By the time it had made its second public appearance, Beriev's Seagull had already established a string of officially recognised international altitude records in the FAI class C3 Group II for turboprop amphibious aircraft. In October 1964, a Be-12, piloted by Capt. M. I. Mikhailov and his crew, attained six records, including an altitude of 39,977 feet without payload, an altitude of 37,290 feet with payloads of 2,205 and 4,409 lb, an altitude of 30,682 feet with a 22,046 lb payload, and an altitude of 6,560 feet with a maximum payload of 22,266 lb. The Be-12 had a maximum speed of 378 mph and a maximum range of 2,485 miles at 199 mph.

By 1965 the Be-12 had become the AV-MF's standard medium-range maritime patrol reconnaissance aircraft and was serving with the Baltic, Black Sea and Northern fleets. It soon built up a reputation that rivalled the Be-6 for being a reliable and rugged flying boat with excellent on-the-water characteristics. Being amphibious, the Be-12 could operate all year round in places such as the Arctic, and under conditions that kept earlier Beriev flying boats restricted to their ice-free bases.

In 1968 it was announced that the Be-12 had broken more FAI records for amphibians, this time for speed. On 24 April, the aircraft established a C3 Group II 310.6-mile closed-circuit speed record of 343 mph, and two days later established a similar record in the C2 Group II for flying boats of 351 mph. The latter was thought to have been established by a specially modified Be-12, which had had the undercarriage removed. However, it was not only in the field of record-breaking that the Be-12 was coming to prominence.

From 1968, the AV-MF operated a detachment of Be-12s from an Egyptian air base for the surveillance of the US Sixth Fleet in the Mediterranean. The aircraft – the first Be-12s to operate outside the Soviet Union – were crewed by Soviet personnel, although the aircraft appeared in Egyptian Air Force camouflage and markings. The detachment continued until 1972 when all Soviet military personnel were expelled by President Anwar Sadat following the Six-Day War with Israel.

By the end of the 1970s, the last of more than 200 Be-12s had been completed at Taganrog, and the type had become the largest amphibian currently in military service anywhere in the world, with the Soviet Union being the only major power to operate flying boats in frontline service. In the meantime, the Be-12 continued to break aviation records, including those for speed over a 62-mile closed circuit in the classes for amphibians, which were set by Capt. G. Yefimov and his crew on 19 April 1976 at 371.2 mph and 370.7 mph respectively, while Capt. V. Svyatoshnyuk established an altitude record of 30,715 feet on 3 May. The Be-12 held no less than forty records – or, rather, all the available FAI turboprop amphibian and flying boat class records – and the last of which was set in 1983.

Not only was it retained by the ASV-MF's Northern Fleet, covering Norway and Sweden, and the Baltic Fleet covering Denmark for coastal patrol, anti-ship, ASW and ELINT duties, but it was also reported that up to a dozen Be-12s were operated from Cam Rahn Bay in Vietnam, from where they patrolled the South China Sea watching for elements of US Navy's Pacific Fleet, which was based in the Philippines. Although the amphibians were wearing Vietnam People's Air Force markings, it was thought that they were flown by Soviet aircrews that had been detached to the Far East, as had been the case in Egypt years earlier.

Beriev's Seagull established more than twenty world records for turboprop amphibians in the late 1960s, during which time more than 200 were delivered to the Soviet Navy's Baltic, Black Sea and Northern fleets, and the BE-12 remains in service with the Russian, Ukrainian and Vietnamese navies to this day.

Since the end of the Cold War, Beriev has converted a small number of surplus Be-12s into scientific research and ecological reconnaissance platforms, as well as into water bombers. Fires in Russia's Far Eastern provinces have been increasing in number and intensity since the 1990s and the country has had a requirement for up to 100 water bombers, but no available budget. As such, the converted Be-12s proved to be a cost-effective interim solution, although their number was limited to fewer than a dozen aircraft.

The Be-12P was modified to scoop 1,500 gallons of water into three internal tanks in fifteen seconds while skimming at 75 mph. Operating 60 miles from a base and 5 miles from available water, the amphibian could drop a total of 40 tons of water before having to refuel.

However, the robust construction of the Be-12PS has seen examples of Russian Navy aircraft being extensively overhauled at Taganrog in 2016 to extend their in-service life for another decade in an SAR role, though they are no longer operated from the water.

It would be another twenty years, and after Georgii Berviev's death in 1979, that his OKB began to develop the Be-12's successor for the Soviet Navy. One of the last, and best-kept Soviet secrets, which was only revealed at the end of the Cold War, was the Beriev A-40 Albatross – the world's most advanced and largest jet-powered amphibious flying boat. Bearing a close resemblance to Saunders-Roe's unbuilt P.208 twin-engine maritime reconnaissance flying boat, with a wingspan of over 136 feet and a length of 143 feet, the A-40 was powered by two 26,450 lb thrust Aviadvigatel D-30KPV turbofan engines that were mounted on short pylons above the wings, plus two 6,500 lb thrust R-60K booster jets that were used for water take-offs. A fully pressurised fuselage accommodated a crew of eight, comprising two pilots, a flight engineer, radio operator, navigator/observer and three mission specialists. With a maximum speed of 500 mph, the amphibian could carry

6.5 tons of ASW weapons over a range of 3,000 miles, which could be extended with in-flight refuelling.

The first of two prototypes of the long-range maritime reconnaissance, anti-submarine warfare and mine-laying A-40 amphibians flew in December 1986 – NATO codenamed 'Mermaid' – and the second flew two years later, with the Soviet Navy placing an order for forty production aircraft, with deliveries to begin in 1993. Unfortunately, as production of this remarkable aircraft began, the Soviet Union imploded and defence funds dried up. The programme was formally cancelled in 2005. Only three A-40s were ever built, one of which was used for static testing.

A projected version was the Be-42, an SAR amphibian with a crew of nine, comprehensive search/weather radars, two LPS-6 semi-rigid rescue boats, a rescue hoist, and a special loading hatch for up to fifty-four survivors or twenty-one stretchers. On board was to be a fully equipped operating theatre with a resuscitation unit and accommodation for extra paramedics. The Be-42 was 90 per cent complete when the programme was suspended due to lack of funds.

The world's first twin-jet engine flying boat was the Russian Beriev R-1, which, powered by two 6,000 lb thrust Klimov VK-1 turbojets, first flew at Taganrog on 30 May 1952. (Beriev)

Seen on its beaching gear, the twin-jet Be-10 was designed to replace the piston-engine Be-6 for maritime reconnaissance, anti-ship attack and anti-submarine warfare missions. (Beriev)

With swept back wings and 14,330 lb thrust AL-7TP turbojets mounted under the wing roots, the Be-10 had a maximum speed of 565 mph. (Beriev)

A contemporary of the Martin SeaMaster, the Be-10 shared its anhedral swept-wing design, with the addition of wingtip-mounted floats. (Beriev)

Armed with two forward-firing NR-23 cannon, and two more in a radar-controlled tail turret, the Be-10 could also carry a 3-ton weapons payload in a bomb bay, which was located after the bottom step. (Beriev)

The Be-10 jet flying boat – NATO codename 'Mallow' – had a crew of three with the pilot under a fighter-style canopy and the navigator in the glazed bow. (Beriev)

Designed to meet a Soviet Naval Aviation (AV-MF) requirement for a multi-role amphibian, the twin-turboprop Be-12 first flew in February 1959. (Author's Collection)

An AV-MF Be-12 – NATO codename 'Mail' – is showing its wide bow chine, high-mounted engines and rear astrodome while on patrol over the Baltic Sea. (Swedish Air Force)

A pair of Be-12 amphibians at an AV-MF base with a good view of the MAD boom and tail wheel assembly. (Author's Collection)

A rare shot of one of the Be-12 'Mails' that were operated in Egyptian Air Force markings to shadow the US Sixth Fleet in the Mediterranean. (Author's Collection)

The deep-hulled, gull-wing Be-12 carried weapons in an internal unpressurised bay with watertight doors aft of the step and under-wing pylons. (Swedish Air Force)

Two Russian Navy Be-12s undergoing maintenance and overhaul of their airframes and 4,190 shp Ivchenko AI-20DM turboprop engines. (David Oliver)

Several former Russian Navy Be-12s were converted into water bombers which were capable of scooping 1,500 gallons of water into internal tanks in fifteen seconds while flying at 75 mph. (David Oliver)

Small numbers of the venerable Beriev Be-12 remain in service with the Ukraine Navy, patrolling the Black Sea. (US Navy)

In 2016, Be-12PS *Yellow 20* was returned to Russian Navy service following a comprehensive rebuild at Beriev's Taganrog facility on the Sea of Azov. (David Oliver)

The prototype Beriev A-40 Albatross was the world's most advanced jet-powered amphibian when it first flew from Taganrog in December 1986. (David Oliver)

Three A-40 Albatross amphibians were built at Taganrog – two flying prototypes and one for static testing. This Be-42 SAR version prototype was 90 per cent complete when the programme was suspended in 1993. (David Oliver)

Designed to take off and land in Sea State 6, the A-40 Albatross – NATO codename 'Mermaid' – had outstanding aerodynamic and hydrodynamic qualities. (David Oliver)

Powered by two 23,830 lb thrust DK-30KPV turbofan engines mounted on short pylons at the rear of the elegantly swept wing, the A-40 was fitted with an air-refuelling probe and wingtip EW pods. (David Oliver)

Flying over the Sea of Azov, the prototype A-40 Albatross shows its two 6,500 lb thrust R-60K booster jets under its main engines, which were used for assisted water take-offs. (David Oliver)

The second prototype A-40 visited the UK to take part in the 1996 Royal International Air Tattoo at RAF Fairford. In 2005, development was abandoned. (Nigel Ish)

CHAPTER FIVE

Canadian Fire-Fighter

The first flight of what was to become the world's most successful post-war commercial amphibious flying boat, the Canadair CL-215, took place on 23 October 1967. Canadair was formed in 1944 and produced 379 amphibious versions of the PBY-5A Catalina, known in Canada as the Canso, thus giving the company a thorough grounding in the complexities of producing amphibious flying boats. The CL-215 was designed specifically to a Canadian government requirement for a modern firefighting water bomber to replace an ageing fleet of diverse ex-Second World War veteran aircraft that were kept busy fighting devastating forest fires in many Canadian provinces during the summer months, which were proving more costly every year to Canada's largest industry and its economy.

The final CL-215 configuration was a conventional twin-engine amphibian with a gross weight of 41,500 lb. The two 2,100 hp Pratt & Whitney R-2800 radial engines were mounted high on top of the wing, above the fuselage, to keep the propellers out of the spray.

It had a wingspan of 93 feet and was capable of scooping 1,200 gallons of water through retractable scoops under the hull into two removable tanks in ten seconds, which were carried internally, skimming the sea or a lake at a speed of approximately 82 mph.

Deliveries of 125 production CL-215s began in June 1969 and terminated in May 1990. Apart from five Canadian provinces, firefighting CL-215s were sold to the governments of France, Greece, Italy, Spain and Turkey – the largest customer, which eventually operated a total of thirty aircraft – as well as the former Yugoslavia. The Royal Thai Navy operates two amphibians in maritime patrol and search and rescue (SAR) roles, while two more were purchased by a Venezuelan mining company as a twenty-six-passenger transport.

However, the success of the rugged water bomber has not disguised the fact that aerial firefighting is a dangerous business. The fires they target from 100 feet are often obscured by thick smoke fanned by strong winds and surrounded by rising heat eddies. Inevitably, more than a dozen aircraft and their crews have been lost while operating in these difficult environments, but the manoeuvrability and strength of the CL-215 has also saved many crews who would have otherwise perished had they been flying in many other types of aircraft.

From the beginning of its development, Canadair had considered the possibility of powering the CL-215 with either the Rolls-Royce Dart or Allison 501 turboprops, but they were rejected due to slow acceleration and high costs. In 1986, however, CL-215 customers were offered a turboprop retrofit to improve performance and extend the life cycle of these specialist amphibians. Some twenty-five CL-215Ts fitted with 2,380 shp Pratt & Whitney Canada PW123AF turboprops had been modified by 1998, while a new upgraded version, the CL-415, was launched by Canadair's new owner, Bombardier Aerospace, in 1993. Also powered by two PW123A turboprops, the CL-415 is fitted with an electronic flight management system, fully powered flight controls and has air conditioning.

Water scooping is a highly effective firefighting technique that requires an aircraft to scoop a large volume of water from a source, such as a river or lake, to drop a mix of water and fire suppressant over the fire. By using the two hydraulically extended and retracted scoops, designed by Fields of Canada, which are mounted behind the planing step, the CL-415 can scoop water from a site that is as small as 6.5 feet deep and 300 feet wide. The amphibian requires 4,400 feet of flyable area to descend from 50 feet of altitude and scoop 1,620 gallons of water during a twelve-second, 1,350 feet-long run on the water at 70 knots. This highly manoeuvrable aircraft allows pilots to navigate around obstacles such as river bends, still in flying mode, while scooping water.

It can drop 6.5 tons of water or fire retardants from modified tanks using a computer-controlled four-compartment drop system. On an average firefighting mission, wherein the site of fire is 6 miles from the water source, the CL-415 is capable of completing nine drops per hour to deliver 14,500 gallons of water. If the designated water site does not have the capacity for a full water load, a partial load will be used and the CL-415 aircraft will make multiple trips.

The firefighter was marketed with Canadair's integrated fire management system, a computer-based information management system that predicts fire danger, occurrence and behaviour, and collates information from fire surveillance sources, be they aerial patrols, ground observers in lookout towers or by remote TV and infrared sensors. Even a two-hour delay between the start of a fire and its detection can be critical in preventing widespread devastation.

The primary difference between the CL-415 and the CL-215T is an EFIS avionics suite. Additional improvements, which first appeared on the CL-215T, include winglets and finlets, higher operating weights, an increased capacity firebombing system and the addition of a foam injection system.

The turboprop's extra power and reverse pitch propellers shortens the take-off distance to 2,650 feet and the landing run to 2,180 feet, and the CL-415 is able to operate for 96 per cent of the time in the Mediterranean and 80 per cent of the time in the Atlantic.

It acquired the name 'Super Scooper' in light of its greatly enhanced performance as a water bomber and fire suppresser.

The CL-415 was also aimed at the military market in maritime, SAR and Special Forces roles. The CL-415 Multi-Purpose (MP) has underwing hardpoints and pylons for long-range fuel tanks, stores and weapons payloads. The turboprop amphibian was also designed to be an efficient people and cargo mover capable of carrying up thirty-five passengers or 6,600 lb of freight over more than 1,000 miles, with fuel reserves. It was designed to quickly climb to an economic cruising altitude to give the CL-415 a maximum cruising speed of 235 mph.

Ninety-five CL-415s were delivered by 2015 to mostly the same CL-215 customers in the firefighting role, namely Canada, France, Greece, Italy and Spain, with the addition of Croatia and Morocco, with Malaysia being the only country to operate the CL-415MP in maritime and SAR roles. The Malaysian aircraft are fitted with the fully integrated Swedish MSS 6000 ST Airborne Systems, which includes an operator console equipped with a highly automated system for controlling sensors and for the presentation, recording and reporting of mission information, digital data and video recording, as well as a high-speed satellite data link. The CL-415MP has nose-mounted search/weather radar, Side-Looking Airborne Radar (SLAR) and a FLIR sensor for target identification and documentation. It also has satellite communications with an integrated presentation of information from the Direction Finder flight management system (FMS), plus a ground station for pre- and post-mission data processing, as well as an Automatic Identification System (AIS) transponder.

The CL-415MP has a wide rear door for missions such as SAR, coastal and border patrol, drug interdiction and environmental monitoring, and retains its firefighting capabilities. Operated by the Malaysian Coast Guard, the CL-415s have been deployed to fight major fires on board ships, including two tankers that collided in the Malacca Straits, one of which was carrying naphtha, and a container ship that caught fire after leaving Port Klang. In 2014 the CL-415MPs conducted extensive search operations in the Malacca Straits for the missing Malaysia Airlines Flight MH370.

Greece ordered two of its ten CL-415s to be configured for the combat search and rescue (CSAR) role, fitted with the MSS 5000 mission equipment, including SLAR, the wing-mounted FLIR SeaFLIR EO/IR sensor, the nose-mounted Honeywell Primus WX660 weather radar, digital cameras and the provision for Have Quick secure radios and rescue beacon receivers, all of which were installed in Sweden by Saab.

To date, eight CL-415s have been lost in accidents.

Viking Air purchased the CL-415 type certificate from Bombardier on 20 June 2016, along with the older CL-215 and CL-215T aircraft. After its acquisition of the business, and based on feedback from current and prospective operators, Viking elected to offer an Enhanced Aerial Firefighter (EAF) option, building on the CL-215T conversion programme by offering additional operator requested enhancements, and it has acquired a number of CL-215 aircraft for conversion to the new CL-415EAF standard.

The prototype Canadair CL-215T, which was retrofitted with two Pratt & Whitney Canada PW100-37 turboprop engines, successfully completed its first flight at Carterville Airport in Quebec on 8 June 1989. (Canadair)

The Bombardier CL-415 amphibian, powered by two 2,380 shp Pratt & Whitney Canada PW123AF turboprops, was a development of the piston-engine CL-215 that first flew in 1967. (David Oliver)

One of fifteen CL-415s delivered to France's Securite Civile is seen scooping 1,350 gallons of water in twelve seconds over a water distance of 4,400 feet. (Canadair)

A CL-415, operated by Italy's Mediterranean Air Service, is seen using its computer-controlled four-door drop system, which optimises drop pattern. (Canadair)

The Spanish Air Force operates five CL-415 amphibians alongside a fleet of fourteen retrofitted CL-215T aircraft. (Canadair)

With twenty-two in service, Italy's Protezione Civile operates the largest fleet of CL-415 water bombers outside of Canada. (Maartin Visser)

Above: The CL-415 Multi-Purpose (MP) has underwing hardpoints and pylons for long-range fuel tanks and stores for maritime and SAR missions. (Canadair)

Left: A 6.5-ton water bomb is dropped from a turboprop-powered CL-415 firefighter aircraft, the first of which was delivered in 1995. (Canadair)

The CL-415 is capable of completing nine drops per hour to deliver 14,500 gallons of water onto a forest fire. (Canadair)

The Moroccan Air Force took delivery of the last Bombardier-built CL-415 amphibian, which was one of five used in a firefighting role. (David Oliver)

CHAPTER SIX

Sea Wings over the Orient

When Shin Meiwa's SS-2 flying boat made its maiden flight on 29 October 1967, it was the first aircraft that the Japanese company had produced. However, the four-turboprop-powered flying boat could trace its origins back to what was arguably the best maritime reconnaissance flying boat to be produced by any nation during the Second World War – the four-engine Kawanishi H8K2, codenamed 'Emily' by the Allies.

The Kawanishi Aircraft Company became the Shin Meiwa Industry Company in 1949, when Japan was permitted to resume aircraft manufacture. Shin Meiwa was awarded a research and development contract for a maritime reconnaissance flying boat by the Japan Maritime Self-Defence Force (JMSDF) in 1960. An ex-JMSDF Grumman UF-2 amphibian was used as a three-quarter scale model for developmental research. Using high-technology hydrodynamic developments pioneered by Shin Meiwa, a large, deep-hulled, high-winged flying boat, bearing more than a passing resemblance to the wartime 'Emily', left the drawing board in 1965. The designer, Dr Shizuo Kikuhara, was also inspired by the design of the Convair Tradewind, which had flown some fifteen years earlier.

The SS-2 was powered by four 2,850 shp General Electric T64-1H1-10 turboprop engines (built under licence in Japan by Ishikawjima-Harima Heavy Industries) that drove Hamilton Standard 63E60-15 three-blade, constant-speed, reversible-pitch propellers. The SS-2 had been designed from the outset as a short take-off and landing (STOL) flying boat, and to this end its 108-foot wingspan, mounted high on the shoulders, bristled with high-lift leading edge slats, spoilers and 'blown' continuous trailing edge flaps. The large T-tail unit also featured leading edge slats with a 'blown' rudder and elevators to improve handling at low speeds. However, it did not feature retractable wing floats.

The deep fuselage with a deep V-shaped single-step planning bottom, with a conventional bow spray chine and novel spray suppressor groove running around the bow to the propeller line, minimised spray by ducting water from the bow to eject it horizontally further along the hullside. The SS-2 featured a large nose radome and a high-mounted cockpit for excellent visibility.

Following a series of lengthy and comprehensive development trials with the JMSDF 51st Flight Test Squadron, the pre-production SS-2 conducted a series of successful take-offs and landings in seas with wave heights of up to 13 feet. In 10-foot waves with a 25-knot headwind, the flying boat became airborne in only twelve seconds and 260 feet of water, proving that it was capable of operating from open sea in the Pacific for 65 per cent of the year.

The JMSDF ordered a total of fourteen production SS-1 anti-submarine warfare (ASW) flying boats, powered by the more powerful 3,060 shp T64 turboprop. They were to be designated PS-1 in JMSDF service, and the first was delivered to the 31st Squadron of the 31st Air Group at Iwakuni on 1 March 1974.

Having been impressed by the high standard of seaworthiness of the flying boat, the JMSDF had made it known that it would be interested in an amphibious search and rescue (SAR) version of the SS-2 to replace its Grumman UF-2 Albatross amphibians. The only major difference between the flying boat and the amphibian design was an additional 11 feet of wide-track retractable undercarriage in place of the flying boats' integral beaching gear. The prototype SS-2A amphibian flew on 16 October 1974 and by the end of the year three amphibians, designated US-1, had formed the nucleus of the 71st Squadron at Iwakuni, which was commissioned in July 1976. At the same time, the JMSDF also approved the purchase of nine additional PS-1 flying boats to form a second unit.

In its ASW form, the PS-1 had a crew of twelve, comprising pilot, co-pilot and flight engineer on the flight deck, while a navigator, radio operator, two sonar operators, radar and MAD operators and a tactical co-ordinator, plus two observers, were accommodated in the tactical compartment aft of the flight deck. On the lower deck was a galley, equipment storage, two main fuel tanks and the main wheel bay. The PS-1 was equipped with electronic countermeasures (ECM) and a magnetic anomaly detector (MAD), plus dipping sonar, while external armament consisted of four homing torpedoes, sonobuoys or four 330 lb anti-submarine bombs carried in two underwing pods that were mounted between each pair of engine nacelles, plus two wingtip launchers for six 5-inch HVAR rockets.

The PS-1 could remain on station for nearly eight hours, cruising at 175 knots at an altitude of 8,000 feet using two engines, and had a maximum endurance of fifteen hours. It was capable of being refuelled in the air or on the sea from a ship.

Although the PS-1 proved to be reliable a relatively mechanically trouble-free, it required a greater degree of skill from the pilot than that required for a landplane in order to master its STOL capabilities when operating from water. To land a 32-ton flying boat in Sea State 5 – rough seas with waves between 8 and 13 feet high, at a speed of only 45 knots – called for not only cool judgment, but a clear understanding of its sophisticated high-lift systems. Although a total of seven of the twenty-three PS-1s built were lost in accidents, not all of them were due to water operations. Some of the worst losses came from aircraft flying overland, including an accident in May 1978 when a PS-1 impacted terrain at Kochi, killing all thirteen crew, as well as in April 1983 during a low-level pass during an airshow, which killed nine crew, and in February 1984, when another was involved in a mid-air collision that resulted in the deaths of all twelve aboard.

The first PS-1 to be delivered was retired in June 1986 having flown 5,400 hours, by which time they were beginning to be replaced by the P-3C Orion ASW landplane. However, with their ability of landing on water to check out either visual or radar sightings, the PS-1 had an advantage over its land-based rivals. This had been graphically illustrated during the extensive searches that took place in Japanese waters for debris and bodies following the shoot-down of a Korean Airlines 747 by Soviet fighters over the Sea of Okhotsk in July 1983, when the JMSDF PS-1s operated alongside US-1s.

By the end of 1986, the 71st Squadron had a total of ten US-1s on strength, which were being produced at the rate of one every two years. In its SAR role, the US-1 normally carried a crew of eight, comprising pilot, co-pilot, flight engineer, navigator, radar operator, radio operator and two observers. The crew could be augmented by up to three air medics or rescue divers. There was seating for twelve stretcher cases and for MEDEVAC flights thirty-six stretchers could be carried, as well as up to five medical attendants.

The US-1 was fitted with a Litton AN/APS-80N nose-mounted search radar and Loran A and C, and Doppler, radar navigation equipment. An HPN-10B wave-measuring height meter provided an accurate indicator of sea conditions when landing on water. A comprehensive rescue kit included a marker launcher, parachute fares, two droppable life-raft containers, an inflatable ramp and a six-man inflatable dinghy. A recue hoist was fitted above the rear rescue hatch, which had a sliding upper door that could be opened even at speeds of up to 138 mph.

A typical SAR mission entailed a transit flight to the search area, which could be up to 1,000 miles, flying between 10,000 and 15,000 feet at a speed of 230 knots, while the search pattern in the area, at which a six-hour loiter was possible at a distance of 690 miles, or two and a half hours at 1,150 miles, was flown at 1,000 feet at a speed of 175 knots using only two engines. For water landing at maximum weight, fifty per cent flap, with boundary layer control (BLC) on, was used to give a landing speed of 64 knots using 950 feet of water. At a lower weight of 80 tons, sixty per cent flap and BLC on would reduce speed and distance to 55 knots and 720 feet.

The last four of the twenty US-1 amphibians that were built, designated US-1As, were powered by the more powerful T64 turboprop, which was rated at 3,490 shp.

The increased power enabled the US-1A to undertake a number of varied roles, including fishery protection patrols, pollution monitoring and ship supply. Stripped of its SAR equipment, the US-1A is capable of carrying up to 115 troops.

It was not until 1996, more than twenty years after the development of the US-1 began, that the company, now renamed ShinMaywa Industries, made the decision to develop the US-1A Kai. Although this had begun as an upgrade of the US-1A, the project was virtually no different from developing a brand-new model, as three stringent engineering requirements were imposed by the JMSDF: improved flight control during take-off and when landing on water; improved environment for casualties during transport; and enhanced open-sea SAR capabilities. Because the primary mission of search and rescue amphibians is to save lives while landing on rough seas, meeting the requirement for landing weight was absolutely essential, and weight reduction posed the greatest difficulty for the development team. This resulted in reducing the maximum take-off/landing weight of 43 tons and the assembly of the prototype No. 1 was completed in April 2003. After engineering and service tests by the Agency, the US-1A Kai was re-designated US-2, and was officially deployed to the 71st Squadron in March 2006.

Powered by four 4,591 shp Roll-Royce AE 2100J turboprops with FADEC and six-blade Dowty propellers, the US-2 has a modified wing with integral fuel tank, composite wing floats, a pressurised upper hull and fly-by-wire (FBW) controls. It is equipped with an electronic flight instrument system (EFIS), Thales OM-100 Ocean Master search radar, Honeywell RE-220 APU and a LHTEC CTS800-based BLC compressor. It also has de-icing abilities for its wing and tailplane leading edges.

The US-2 has a mission radius of 1,150 miles and a maximum range of nearly 3,000 miles, cruising at 230 knots at 10,000 feet, and can land in only 1,000 feet of water. The unit currently has six US-2s on strength plus a single US-1A that will be replaced by another US-2 within the next two years. In April 2015 a US-2 suffered substantial damage while training in the Pacific Ocean, about 40 km north-east of Ashizuri Cape, Kochi Prefecture. Four of the nineteen crew were injured.

ShinMaywa Industries has yet to secure a foreign customer for the US-2. In December 2015, Japan signed an agreement on the transfer of military technologies with India and discussions on the US-2 for the Indian Navy and Coast Guard have been ongoing since then. Unit cost appears to be the principle obstacle, although Japan has reportedly agreed to a 10 per cent decrease to $113 million for twelve to eighteen aircraft. Deliveries of the US-2s will not only be useful from a military point of view, but will also support the Indo-Japanese bloc in response to China, which has been increasing its presence in the Indian Ocean and is a long-standing ally of India's arch-enemy, Pakistan.

Japan's mainland neighbour, the People's Republic of China, surprised the aviation world when it announced in 1986 that it had designed and built a large maritime reconnaissance and SAR flying boat to replace its fleet of Soviet-era piston-engine Beriev Be-6 flying boats. The Harbin SH-5's deep hull, which resembled the Shin Meiwa PS-1, has a glazed nose surrounding the large radome and twin fins similar to the Be-12. Powered by four 4,200 shp Woijang WJ-5 turboprops – Chinese copies of the Soviet Ivchenco AI-20 engines that power the Be-12 – the SH-5 has a shoulder-mounted wingspan of 118 feet and a length of 127 feet, making it at that point the largest post-war amphibious flying boat in the world. It has a Doppler search radar fitted in a thimble radome in the nose and a fixed MAD boom extended from the tail section. It is also fitted with fixed-wing floats, which are attached to V struts, and has a Shin Meiwa-inspired single-wheel beaching gear.

The SH-5 carries a flight crew of eight, including a pilot, co-pilot, navigator, flight engineer, radio operator and three systems specialists. The number of specialists may vary depending on the mission. The fuselage is unpressurised and it has three freight compartments behind the cockpit area, followed by a cabin for the mission crew, a compartment for communications gear and, finally, the stores compartment. There is a corridor connecting all three compartments, with watertight doors between compartments, and there is one crew door on the left side of the aircraft and two doors on the right.

The SH-5 is armed with a dorsal turret-mounted twin cannon, and has two stores pylons under each wing – one placed between the hull and the inboard engine, while the second is placed between the inboard and outboard engines. These four pylons can carry anti-ship missiles or homing torpedoes. A total of 6 tons of other stores, such as depth charges, mines, bombs, sonobuoys, or rescue gear, can be stored in a rear compartment in the fuselage. If configured as a cargo lifter, the SH-5 can carry 10 tons of cargo.

After a protracted design and development that began in 1970, only six production aircraft were built, which entered service with the People's Liberation Army Navy (PLAN) in 1986. Three of them continue to operate from an aircraft base near Qingdao in Shandong Province in 2017. The prototype SH-5 was not delivered to the PLAN and was converted into a water bomber for the aerial firefighting role.

The SH-5's replacement may be the China Aviation Industry General Aircraft (CAIGA) AG600, reportedly the largest amphibious aircraft currently being built. CAIGA, which started work on the prototype AG600 in 2013, has announced that the aircraft will have a maximum take-off weight of 53.5 tons and a maximum range of over 3,000 miles. Powered by four 5,103 shp WJ-6 turboprops – a license-built copy of the Russian Ivchenko AI-20 engine that powers the Be-12, and which drives six-blade propellers – the aircraft has a wingspan of 127 feet and a length of 121 feet. The prototype was rolled out on 23 July 2016 at the Zhuhai AVIC factory.

The Aviation Industry Corporation of China (AVIC) has stated that the AG600 will be suitable for aerial firefighting, being capable of dropping 12 tons of water, as well as for SAR operations for up to fifty passengers while operating in Sea State 5. Sources also note that the aircraft could also have strategic value in the South China Sea, which has been subject to various territorial disputes. Commentators in China claim that the aircraft was specifically designed to defend China's interests in the South China Sea area.

The manufacturer has indicated that they expect export sales of the aircraft and that island countries, including New Zealand and Malaysia, have expressed interest. The AG600 amphibious aircraft has been successfully glide-tested and made its maiden flight in Zhuhai ahead of schedule on 29 April 2017.

The Shin Meiwa SS-2, the prototype of the Japan Maritime Self-Defence Force (JMSDF) PS-1 ASW flying boat, shows its high-lift leading edge slats and MAD stinger, which is mounted at the top of its 'T' tailplane. (Shin Meiwa)

A JMSDF Shin Meiwa PS-1 ASW flying boat is seen standing on its integral beaching gear at the 31st Air Group's base at Iwakuni. (J. J. Halley)

Six PS-1 flying boats belonging to the JMSDF's 71st Squadron of the 31st Air Group.

The Shin Meiwa US-1 SAR amphibian flew for the first time in October 1974 and was assigned to the JMSDF's 51st Operational Training Squadron in 1975. (JMSDF)

Serving with the 71st Squadron, this JMSDF Shin Meiwa US-1 amphibian has its 'blown' flaps extended. (A. Heape)

A 71st Squadron US-1A tests its uprated Ishikawajima-built General Electric T-64-IHI-10J turboprops at Gifu. (A. Heape)

The rescue crews of this 71st Squadron US-1A are seen practicing with the amphibian's outboard, motor-powered, inflatable dinghy. (JMSDF)

The upgraded ShinMaywa US-2, powered by four Roll-Royce AE 2100J turboprops with six-blade propellers, first entered service with the 71st Squadron in 2006. (JMSDF)

This JMSDF US-2 STOL SAR amphibian shows its distinctive bow chine, main wheel bulges and domed observation windows as it takes off. (JMSDF)

A JMSDF US-2 SAR amphibian, operated by the 31st Fleet Air Wing, is seen taxiing near Hansin Air Base on Honshu. (Mamo)

A 71st Fleet Squadron US-2, showing its six-blade propellers and twin-main wheel undercarriage, is seen taxiing at Tokyo Haneda International Airport. (Aeroprints.com)

The prototype of the Chinese People's Liberation Army Navy's Harbin SH-5 maritime patrol flying boat was converted to a water bomber. (CATIC)

One of six SH-5 maritime patrol flying boats delivered to the Chinese People's Liberation Army Navy (PLAN) in the 1970s is seen on the step. (Author's collection)

The deep-hulled PLAN SH-5 is equipped with a magnetic anomaly detection (MAD) boom extending from the rear fuselage, as well as fixed wing floats and single-wheel beaching gear. (Author's collection)

The four-turboprop engine, 53.5-ton AG600 amphibian features a spray-suppressor groove running around the bow and a high 'T' tailplane similar to that of the ShinMaywa US-2. (Alan Warnes)

The China Aviation Industry General Aircraft AG600 amphibian, designed for maritime patrol, SAR and firefighting roles, made its maiden flight on 29 April 2017. (Alert5)

CHAPTER SEVEN

Back to Beriev

After the disappointment of the cancellation of the A-40, and the virtual halt of military contracts, Beriev's chief designer, Gennadi Panatov, went back to the basics. However, the technology used in these advanced jet amphibians has not been entirely lost. In 1990 Panatov formed the Taganrog Aviation Scientific Technical Komplex (TANTK) to design, manufacture and market a series of commercial flying boat designs, one of which was a scaled-down A-40 to meet the civil market – the Be-200. In 1990, the Russian government approved a purpose-designed water bomber version of the new design, which was to be commercially developed by the Beta Air consortium (made up of the Beriev OKB, the Irkutsk Aircraft Production Association, the Ukrainian Bank Prominvest and ILTA Trade Finance S. A. of Geneva).

Beriev was responsible for the design, development and the static and flight-testing of the two prototypes, which were to be built at Irkutsk in Siberia. The Be-200 made its maiden flight from Irkutsk on 24 September 1998 – the first new Russian aircraft to fly since the end of the Cold War – after which it was ferried to Taganrog to begin water operations, certification trials and water-drop test flights. These tests proved that the Be-200 would be in a class of its own as a water bomber.

Powered by two 16,550 lb thrust ZMKB Progress D-436TP turbofan engines, the 107.5-foot wingspan, 37-ton amphibian has a maximum speed of 445 mph and a maximum range of 2,400 miles. It has the same supercritical wing design with leading edge slats and speed brakes as the A-40, with the same 'T' tail configuration and position of the engines, which have also been mounted on pylons above the wings.

The Be-200's integrated logistics support system, the ARIA-2000, which was jointly developed by the Russian Aviation Equipment Institute and the US company AlliedSignal, now Honeywell, enables the two-man crew to compute the speed and distance required to fill its water tanks, as well as the speed, distance and height, down to 150 feet, to make the most effective use of its 12-ton water 'bomb' over a forest or industrial fire. The water drops can be carried out singly or sequentially, depending on the characteristics of the fire. In its firefighting support role, the Be-200 can also accommodate up to twenty-six firefighting personnel and their equipment.

President Boris Yeltsin established the Russian Federation Ministry of Emergency Situations (EMERCOM) on 10 January 1994. It was tasked with managing the Civil Defence and search and rescue (SAR) services in the Russian Federation and directing activities aimed at eliminating the consequences of large-scale natural disasters and other civil emergencies. Currently, EMERCOM operates more than seventy aircraft, including Beriev Be-200 amphibious firefighting aircraft.

As part of the EMERCOM 2030 programme, additional indigenous aircraft will be delivered to the ministry and a second batch of the Be-200 twin-jet amphibious aircraft has been ordered. The Be-200 firefighter suppresses fires by dropping water and/or chemical retardants. Eight ferric aluminium alloy water tanks are located under the cabin floor in the centre fuselage section, which can be filled as the aircraft skims rivers, lakes or the open sea at 160 km/h. Four retractable water scoops – two forward and two aft of the fuselage step – can be used to scoop a total of 12 tonnes of water in fourteen seconds. Alternatively, the tanks can be filled from a hydrant or a water cistern on the ground. Up to thirty-six fire fighters or 7 tons of firefighting equipment can be carried, or up to thirty fully equipped smoke jumpers can be seated along the side walls of the cabin, which has a jump door at the rear on the starboard side.

Water can be dropped in a single salvo, or in up to eight consecutive drops. The Be-200 can also be fitted with six internal auxiliary tanks for fire-retarding chemical agents, with a total capacity of 1.2 m³. The aircraft can empty its water tanks over the site of a fire in 0.8 to 1 seconds when flying above the minimum drop speed of 120 knots (135 mph). The Be-200 is capable of taking off from either a 4,100-foot-long land runway or a stretch of open water no less than 3,300 feet long and 8 feet deep in Sea State 3 with waves that are up to 5 feet high.

In its SAR role, the amphibian is equipped with an Orion 25S inflatable boat, naval radios, a 600 W loudspeaker and searchlight, storage for twenty inflatable rafts, two four-seat motorboats and an observer's station. It can be fitted with fifty-seven folding seats and has attachments for thirty litters.

Now part of Russia's United Aircraft Corporation (UAC), Beriev is focusing on the production of the latest variant of the firefighting amphibian – the Be-200ES. 'The plane has undergone a major upgrade, with more than 60 per cent of design changes,' said Yuri Slyusar, the President of UAC. The aircraft is being supplied to the domestic market under contracts for the Russian Ministry of Defence, which has ordered six in an SAR role, and for EMERCOM, which has ordered six Be-200ES aircraft in addition to five Be-200ChS Irkut-built aircraft that have already been delivered – two of which were delivered in 2016. A single Be-200ChS has been exported to the Azerbaijan Ministry of Emergencies. The current general director of the Beriev Aviation Company, Yuri Grudinin, claims that the Be-200ES has good export potential because it is the most modern machine in its class and is the only amphibious jet aircraft in the world. It also has an advantage in the speed of extinguishing fires and it can carry twice as much water or chemical retardants as the nearest competitor.

The Be-200ES features an upgraded fly-by-wire (FBW) electronic flight control system and a redesigned centre section and wings, based on the experience of operating the firefighting aircraft. The designers have modified the flight-navigation system and also improved the hydraulic system. Radio communication equipment has been replaced by more modern systems and other systems have been replaced or upgraded, in particular the weather radar and thermal detection systems. Beriev is currently producing eight Be-200s per year at its Taganrog plant. The new amphibians are tested and certified at the Beriev Experimental Test Base at Gelenzhik, located on the coast of the Black Sea, north of Sochi, which is also an EMERCOM operational centre.

During the spate of serious forest fires that have plagued Europe over the past few years, several governments have contracted the Be-200s, and they have also been evaluated in Croatia, France, Greece, Israel, Italy and Portugal, as well as the Far East and North and South America. Once again, a single example was sold to the Azerbaijan Ministry of Emergences in 2008. During the 2017 International Paris Air Show, Beriev Aircraft and the Argentine Ministry of Defence's Secretariat for Logistics and Coordination in Emergency Situations signed a Memorandum of Understanding (MoU), which expressed intentions of delivering three Be-200ES amphibians, as well as the creation of a service centre and maintenance, repair and overhaul (MRO) facilities for Be-200 aircraft in the Argentine Republic. After a series of disastrous forest fires that destroyed homes and industrial facilities in 2016, the Chilean Air Force officially confirmed its interest in acquiring up to three Be-200 firefighting aircraft in July 2017, which would have a secondary search and rescue and community support role in seasons when fire danger is lower. Also announced at the 2017 Paris Air Show was an agreement with PowerJet for the possibility of re-engining the Be-200 with 15,000 to 18,000 lb thrust SaM146 turbojet engines. The SaM146 is produced by the PowerJet joint venture between Safran of France and NPO Saturn of Russia. Another contract between Beriev Aircraft and the Chinese company Energy Leader Aircraft Manufacturing Co. Ltd was signed at Taganrog in June 2017 for the delivery of two Be-200ES amphibians, with an option for two more to the People's Republic of China.

Among other variants on offer is a sixty-four-passenger version that can be reconfigured to carry 16,500 lb cargo or a combination of 6,600 lb of cargo with twenty-eight passengers. An air ambulance could accommodate thirty stretchers plus seven casualties or medical staff, and a search and rescue version is capable of loitering at 200 nautical miles from its land base for up to six and a half hours. The Be-210 is a proposed development, with a strengthened airframe to accommodate seventy-two passengers and two flight attendants, as well as extra fuel to enable it to fly 1,000 miles with a full load of passengers, for use with airlines in regions of minimal airport infrastructure. In a military role, the Be-220 is a maritime reconnaissance variant, coming equipped with a Novella Sea Dragon detection system and an extended nose radome. It was offered to the Chinese Navy and is a contender for an Indian Coast Guard competition. If it is successful, the Beriev flying boat will have come full circle since the Be-6 first entered service with the Soviet Navy sixty-five years ago.

The full-scale mock-up of the Beriev Be-200, a jet-powered amphibian aimed at the civil market, was constructed at Taganrog in 1991. (David Oliver)

The two Be-200 prototypes and six production amphibians were produced at the Irkut Corporation plant at Irkutsk, in Siberia. (David Oliver)

The first prototype of the jet-powered Be-200 is seen at Irkutsk on 17 October 1998, when it made its first official flight before transferring to Taganrog to begin certification trials. (David Oliver)

Small, wedge-shaped hydrodynamic compensators are located aft of the Be-200's step to aid 'unsticking' from water in wave heights of up to 4 feet. (David Oliver)

The second Be-200 prototype is seen on the ramp of the Beriev Experimental Test Base at Gelenzhik, which is located on the coast of the Black Sea. (David Oliver)

A Be-200 water taxiing in Gelenzhik Bay, where the new production Beriev amphibians are tested and certified. (David Oliver)

The Be-200 on the step during its 3,300-foot take-off run, illustrating how its high-mounted jet engines and 'T' tailplane are clear of the spray. (David Oliver)

A Ministry of Emergency Situations (EMERCOM) Be-200ChS firefighter comes ashore and climbs the test base ramp at Gelenzhik under its own power. (Beriev)

The third EMERCOM Be-200ChS demonstrates its capability of dropping a 12-ton water bomb in a single three-second drop. (Beriev)

A factory-fresh Be-200ES – the latest upgraded firefighter, which features an upgraded fly-by-wire (FBW) electronic flight control system and a redesigned centre section and wings – is seen at the Gelenzhik test base. (David Oliver)

The first of six upgraded Be-200ES firefighting amphibians ordered for EMERCOM service was delivered in January 2017. (UAC)

A Beriev Be-200 gives a patriotic demonstration of the amphibian's sequential drop capability using different coloured liquids. (Beriev)

CHAPTER EIGHT

The Flying Boat's Fragile Future

Now well into the twenty-first century, there are a few hundred turboprop and jet-powered flying boats in operation – their primary role being that of fighting fires followed by search and rescue and maritime reconnaissance. Japan and Russia continue to build new aircraft – albeit at a very slow rate, which makes them expensive – and they are being joined by China.

Over twenty years ago, ShinMaywa, in conjunction with the Society of Japanese Aerospace Companies, carried out long-term studies into the possibility of developing new transport systems using amphibious aircraft. ShinMaywa produced a number of new flying boat concepts that could have become an integral part of such a system, including a thirty to fifty-passenger twin-jet-powered amphibious feeder-liner, 250- and 400-seat amphibians powered by four turbojets, and a giant 1,200-seat long-haul commercial flying boat.

The most advanced, and possibly the most practical of these projects was the SS-X amphibious feeder-liner, which reached the point of wind tunnel testing. It would have been an STOL amphibian, with a maximum take-off weight of 17.5 tons and capable of carrying forty passengers at a cruising speed of 350 mph over a range of 500 miles. It was powered by two advanced 8,000 shp turbofan engines, which were positioned high above the wing, and which utilised upper-surface blown flaps, spoilers and high-lift leading edge slats. The tailplane, similar to that of the US-2, was T-mounted to be clear of the jet exhaust and water spray, while the wing floats would be non-retractable to save both weight and cost.

However, although the SS-X remained a project, Beriev released details of an almost identical design at the same time – the Be-114.

A versatile, amphibious commuter aircraft intended for transportation of up to forty-four passengers and up to 6 tons of cargo in a pressurised cabin, the Be-114 was designed with a rear cargo ramp, which provides the easy loading of variable cargoes, including vehicles and large containers. Similar to the SS-X, it has engines mounted above the high-wing, T-mounted tailplane and fixed wing floats, but is instead powered by two 2,800 shp TV7-117-2 turboprops. The Be-114 would have a maximum take-off weight of 17.5 tons, a cruising speed of 330 mph and, with its maximum payload, a range of 650 miles.

Although details of the Be-114 project were released more than a decade ago, in 2016, the Beriev Aircraft Company's director general, Yuri V. Grudinin, claimed that design and development of the amphibian has been accomplished and production could begin as soon a funding was made available.

Having launched the AG-600 programme, China is also heavily involved in the development and production of one of the most transient of any recent flying boat projects – the Dornier Seastar.

A lifelong flying boat pioneer, Claudius Dornier's company is still involved in flying boat projects today. In 1981 Dornier won a contract from the German Federal Ministry of Research and Technology to study a new flying boat design. As a research prototype, the company converted a pre-war-designed ex-Spanish Do 24 flying boat as a technology demonstrator. Three 1,125 shp Pratt & Whitney Canada PT6 turboprops replaced the original BMW piston engines and a radical new wing design and tricycle undercarriage was fitted. The 'new' Do 24TT first flew in April 1983 and successfully completed 85 hours of flight and sea trials over the next two years, but the project was not developed any further.

At the 1983 Paris Air Show, Claudius Dornier came out of retirement to announce his latest design, the Seastar, a high-tech amphibian inspired by his first successful flying boat, the Wal. With a metal wing design that mirrored that of the Do 24TT, the fuselage was built of composite materials. The ten-seat Seastar was powered by two 500 shp PT6A turboprops mounted in tandem above the wing and featured the Dornier trademark sponsons in place of conventional wing floats. The first prototype made its maiden flight in August 1984 but was badly damaged when landing on Lake Constance with its wheels down a year later.

In April 1986 Claudius Dornier died and the project was taken over by his two sons, and although a second prototype had flown, the project faltered due to lack of investment. Following liquidation of the Dornier Composite Aircraft Company in 1991, a reformed Seastar company was formed as a joint venture with a Malaysian group. The new venture planned to manufacture the Seastar on a site near Penang's International Airport at the end of 1996; however, the economic crisis of the late 1990s hit the Far East badly and the plan came to nothing.

The Seastar programme was re-launched in 2009 with the formation of the Dornier Seaplane Company in Canada with Conrado Dornier as its chairman, but it again failed when backers withdrew their support.

However, Dornier Seawings GmbH was formed in 2013 as a joint venture between the Dornier family and two state-owned Chinese enterprises, with the former providing the necessary design and certification and the Chinese partners providing the financial backing to ensure production and a sustainable business model. Production lines are currently in Germany at Oberpfaffenhofen and ultimately at Wuxi in China.

Production aircraft will be based on the all-composite constructed Seastar CD2, which is powered by two 650 shp Pratt & Whiney Canada PT6A-135A turboprops driving four-blade propellers. With a maximum cruising speed of 180 knots and a maximum range of 900 miles, the Seastar CD2 has a take-off run of 1,850 feet on land and 2,500 feet on water, with a landing distance of 2,250 feet on land and 2,480 feet on water.

The twelve-seat Seastar CD2 is being offered as a multi-purpose aircraft, capable of delivering unmatched versatility at low operating costs for a wide range of missions, from VIP-transport to commercial, corporate and governmental/special missions. In an air ambulance configuration, the Seastar can accommodate up to three patient stretchers with associated medical equipment.

The Seastar could be an ideal aircraft for operations such as coastal surveillance, patrolling, environmental control, fisheries protection, SAR, drug interdiction and disaster relief. It would have the ability to combine both surveillance and immediate intervention by landing on water wherever necessary. Operating in single-engine mode extends on-station endurance up to nine and a half hours.

The production Seastar cockpit will feature Honeywell's Primus Epic 2.0 avionic suite with advanced vision, communication, navigation, surveillance and air traffic management systems, which allows for single-pilot operation.

In 2016 Dornier Seawings GmbH selected Diamond Aircraft Industries Inc. in Canada to produce the Seastar's all-composite airframe. The initial contract – for the manufacture of the first ten ship sets, with options for subsequent units – includes significant tooling work to ready the Seastar for higher volume production. Delivery of the first airframe to the Dornier Seawings facilities in Oberpfaffenhofen for final assembly and completion was planned for mid-2016. However, problems with productionising the aircraft led to delays for the amphibian's first flight.

Dornier Seawings is scheduled to receive the amended Type Certificate by the European Aviation Safety Agency (EASA) in the second quarter of 2018 and will deliver the first aircraft to customers later that year.

We will have to wait until then in the hope that the protracted life cycle of the innovative Seastar fulfils the promise it deserves.

The facts are that 70 per cent of the world's surface is covered by water, 80 per cent of global economic activity takes place within 150 miles of the oceans and most of the world's capital cities are situated on the coast or on the banks of large rivers. Water runways cannot be destroyed by natural disasters or bombs, and it is clear that there will always be a demand of the flying boat as long as aircraft are used for private, commercial or military purposes. In spite of earlier predictions, technology will hopefully ensure the preservation of the species.

The several amphibious commuter aircraft projects that Beriev has had on the back burner include the innovative twin-turboprop, forty-four-seat Be-114, which was designed with a rear cargo ramp. (Beriev)

As a flying boat research project, Dornier re-fitted a pre-war Do 24 into a turboprop-powered amphibian in 1983. (Dornier)

The Dornier 24 Technology Testbed (TT) was fitted with a new high-lift wing, three Pratt & Whitney Canada PT6 turboprops and extended sponsons. (David Oliver)

Claudius Dornier's son, Iren, flew the Do-24TT on a world tour in 2004 in support of the United Nations Children's Fund (UNICEF). (David Oliver)

The first prototype of the composite-hull Claudius Dornier Seastar amphibian made its maiden flight from Lake Constance on 17 August 1984. (Claudius Dornier)

Powered by two Pratt & Whitney Canada turboprops that were fitted in tandem, the original ten-seat Seastar was badly damaged when it landed on Lake Constance with its wheels down. (Claudius Dornier)

The Seastar's high-lift wing design and extended sponsons, into which the main wheel retracted, were developed from the Do 24TT. (Claudius Dornier)

The first all-composite twelve-seat Seastar CD2 was flown for the first time in April 1987, but was grounded following liquidation of the Dornier Composite Aircraft Company in 1991. (David Oliver)

A second pre-production Seastar CD2 featured an enlarged cockpit, larger cabin windows, a re-profiled nose and four-blade propellers, while the wing struts were omitted altogether. (Dornier Seawings)

Production and marketing of the thirty-year-old Seastar project is currently undertaken by Dornier Seawings GmbH, a joint venture that was formed in 2013 between the Dornier family and state-owned Chinese enterprises. (Dornier Seawings)